LEGUME INOCULANTS AND THEIR USE

A pocket manual jointly prepared by

Nitrogen Fixation for Tropical Agricultural Legumes (NifTAL) Project, USA

and

FAO Fertilizer and Plant Nutrition Service

Land and Water Development Division

in association with

FAO Crop and Grassland Production Service

FOOD AND AGRICULTURE ORGANIZATION OF THE UNITED NATIONS
Rome, 1984

The designations employed and the presentation of material in this publication do not imply the expression of any opinion whatsoever on the part of the Food and Agriculture Organization of the United Nations concerning the legal status of any country, territory, city or area or of its authorities, or concerning the delimitation of its frontiers or boundaries.

P-52
ISBN 92-5-101441-8

Printed in Italy

PREFACE

The increased use of symbiotic fixation of nitrogen by *Rhizobium* bacteria should help in achieving increased yields of food and forage legume crops in a most economical way.

The need to disseminate knowledge on the possibilities offered and on appropriate techniques has prompted the specialized technical services of the Food and Agriculture Organization of the United Nations and the Nitrogen Fixation for Tropical Agricultural Legumes Project (NifTAL), USA, to join forces for the production of this manual. The basic text was prepared by Dr Joe Burton of NifTAL, who also contributed illustrations.

It is hoped that agricultural extension personnel, progressive farmers and agencies engaged in encouraging legume grain and fodder production will find this manual useful.

The handbook may be republished locally in any member country of FAO, with acknowledgement to FAO and NifTAL, either in one of the official languages of FAO or in the local vernacular. In such cases, this preface may be rewritten with special reference to local conditions, and a supplementary text giving reference to local conditions can be added.

Comments and suggestions for effecting further improvements in the manual would be most welcome.

CONTENTS

Figures

Page

Tables *Page*

1. NITROGEN-FIXING LEGUMES

Legumes (species of Leguminosae) are among the world's most important crops. They provide nutritious body-building food for man and animals the world over. In addition to being rich in protein, leguminous crops are also high in bone-building minerals and vitamins essential to good health.

Leguminous plants can obtain most of the nitrogen they need from the vast supply of gaseous nitrogen in the air. Air is 80 percent nitrogen; there are around 6 400 kilograms of nitrogen above every hectare of land and water. Leguminous plants gather and use this nitrogen by working symbiotically with special bacteria (rhizobia) in nodules on their roots. Rhizobia infect the root hairs of the leguminous host; nodules develop and become small nitrogen factories on the legume roots (Fig. 1).

The host plant provides a home for the bacteria and energy to fix or gather air nitrogen (N_2). In return, the plant receives fixed nitrogen from the nodule and produces food and forage protein. Legumes also leave fixed nitrogen in the soil for succeeding crops. Since nitrogen is commonly the most limiting element in food production, and one of the most expensive in fertilizer, this special ability of leguminous crops to work symbiotically with rhizobia to produce protein is becoming increasingly important in world agriculture.

The family Leguminosae comprises around 20 000 plant species in about 650 different genera. Only about 15 percent of these species have been studied. Many of our

Figure 1. Well-nodulated bean plant (*Phaseolus vulgaris*)

most important food crops belong to this family. The economic success of multibillion dollar crops such as soybeans and groundnuts depends upon symbiotic nitrogen fixation by these legumes in association with rhizobia in nodules on their roots.

World-wide, it is estimated that leguminous plants gather or fix 80 million tons of nitrogen annually from the vast free supply in the air. Crop legumes account for around 35 million, and others, in meadows, grassland and forests, for about 45 million tons. In contrast, fertilizer manufacturers produce at high cost only 50 to 60 million tons of nitrogen annually. The amounts of nitrogen fixed by some important legume-*Rhizobium* associations are shown in Figure 2. This varies widely, depending on the legume, the strain of nodule bacteria and the soil. A list of some legumes and reported amounts of nitrogen fixed is given in Table 1 and Figure 2.

The current energy crisis has driven up the price of mineral fertilizer and developing countries lack the foreign exchange required to import it, so the role of legume-fixed nitrogen will have to be greatly expanded if world demands for food and forage are to be met.

2. THE NITROGEN CYCLE

Nitrogen is the essence of all life, both plant and animal. It is inert, meaning it is difficult to combine chemically with other elements. Nitrogen compounds are continually being changed or broken down as a result of biological or chemical action. Figure 3 illustrates some of these changes. Elemental or gaseous nitrogen may be fixed or combined with other elements by bacteria (*Rhizobium* sp. and *Frankia* sp.) working in nodules, by blue-green cyanobacteria working symbiotically with *Azolla*, a small fern, and by some associative and free-living bacteria. Another way to change

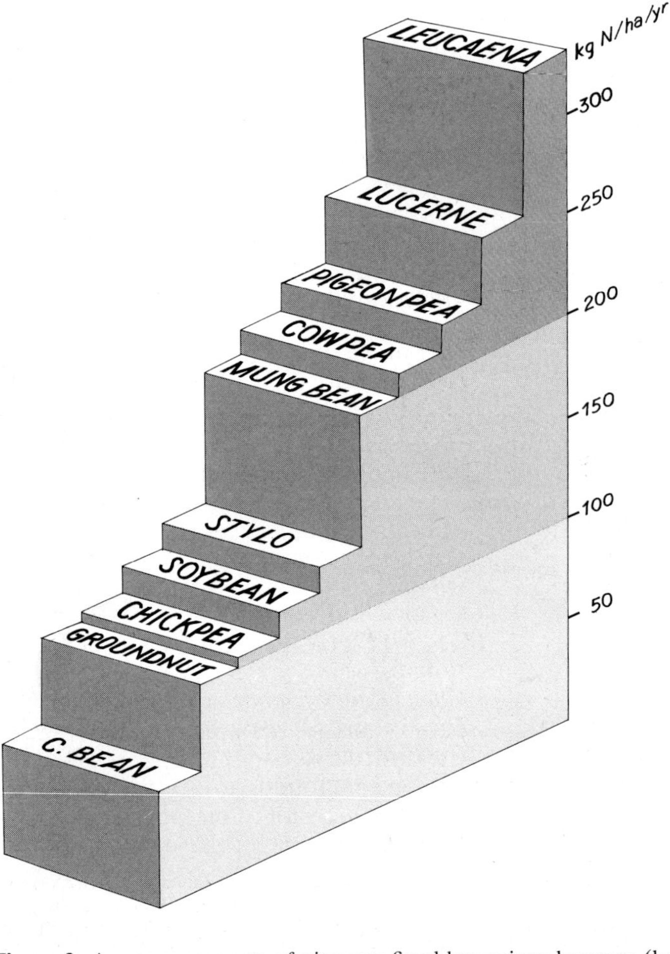

Figure 2. Average amounts of nitrogen fixed by various legumes (kg N/ha/yr)

Table 1. ESTIMATED AMOUNTS OF NITROGEN FIXED BY VARIOUS LEGUME CROPS UNDER FIELD CONDITIONS [1]

Plant		Nitrogen fixed (kg N/ha/yr)
Food legumes		
Calapo	*Calopogonium mucunoides*	370-450
Horse bean	*Vicia faba*	45-552
Pigeon pea	*Cajanus cajan*	168-280
Cowpea	*Vigna unguiculata*	73-354
Mung bean	*Vigna mungo*	63-342
Guar	*Cyanopsis tetragonoloba*	41-220
Soybean	*Glycine max*	60-168
Chick-pea	*Cicer arietinum*	103
Lentil	*Lens esculenta*	88-114
Groundnut	*Arachis hypogaea*	72-124
Pea	*Pisum sativum*	52-77
Bean	*Phaseolus vulgaris*	40-70
Forage legumes		
Tick clover	*Desmodium intortum*	897
Sesbania	*Sesbania cannabina*	542
Leucaena	*Leucaena leucocephala*	74-584
Centro	*Centrosema pubescens*	126-398
Alfalfa	*Medicago sativa*	229-290
Subclover	*Trifolium subterraneum*	207
Ladino clover	*Trifolium repens* var. *gigantea*	165-189
White clover	*Trifolium repens*	128
Stylo	*Stylosanthes* spp.	34-220
Vetch	*Vicia villosa*	110
Puero	*Pueraria phaseoloides*	99

[1] The amount of N_2 fixed by legumes varies widely with host genotype, *Rhizobium* efficiency, soil and climatic conditions and, of course, methodology used in assessing fixation. The data here are a composite from two recent reports: La Rue and Patterson (1981), *Advances in Agron.*, 34: 15-36 and P.S. Nutman (1981) Hannaford Lecture, WAITE Agricultural Research Inst., Australia.

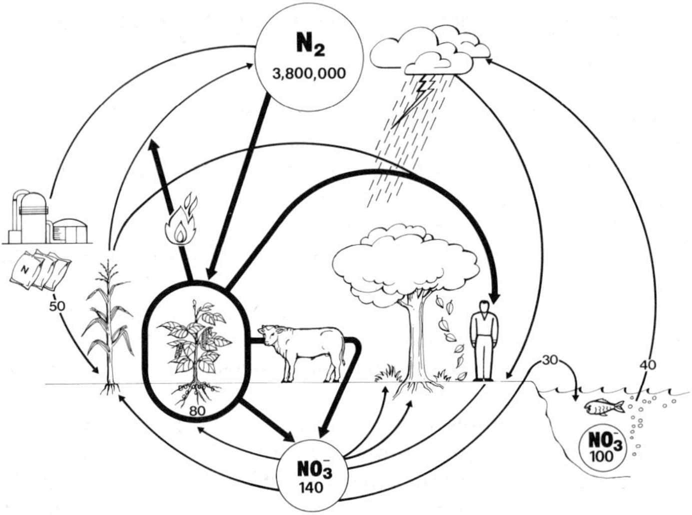

Figure 3. Place of legumes in the nitrogen cycle (10^6 t)

nitrogen is chemically, using a catalyst and fossil fuel to provide the high pressure and temperature necessary for reaction. This process requires much energy. Lightning discharges in the air also bring about some natural nitrogen fixation.

Nitrogen taken from the air can be built into plant protein, which is eaten by animals and converted to body protein or meat for human food. Some of the nitrogen from the animal is eliminated as manure and may be returned to the soil.

Nitrogen added to the soil may be used by other crops or it may be lost through drainage or run-off. Also, certain bacteria may convert soil and plant nitrogen back to gas and it may return to the air again. These changes that nitrogen undergoes are important. Management should always aim to conserve fixed nitrogen for use in crop production.

3. RHIZOBIA — NODULE BACTERIA

Rhizobia are soil bacteria characterized by their unique ability to infect root hairs of legumes and induce effective nitrogen-fixing nodules to form on the roots. Rhizobia commonly occur in soils but often fail to produce effective nodulation, either because too few are present or because those present cannot work effectively with the particular legume.

Rhizobia are rod-shaped living plants which exist only in the vegetative state. Unlike many other soil microorganisms, rhizobia produce no spores. They are aerobic and motile. They multiply by simple cell division. The generation time ranges from two to four hours for so-called "fast-growers", which generally form relatively large (2- to 4-mm diameter) colonies in three to five days, to six to eight hours for the "slow-growers", which yield colonies 1 mm in diameter in seven to ten days. The optimum growth temperature is 28 to 30 °C.

Rhizobia are not particularly fastidious in their nutritional requirements. They can use sugars, alcohols and some acids as sources of energy. Yeast extract provides growth factors and vitamins and usually enhances growth, but some species can produce their own growth factors.

Nodulation

Infection of root hairs

Steps in the infection and development of a nodule are illustrated in Figure 4. The mechanism of recognition between rhizobia and the leguminous seedling is not fully understood, but when an infective *Rhizobium* cell comes in contact with the root of a susceptible legume seedling, the *Rhizobium* cell increases in number and root hair colonization occurs. The root hair curls; a *Rhizobium* enters, mul-

infection thread cells in this zone multiply actively and are contaminated with *Rhizobium*, but fixation does not take place

meristem made up of small cells that are not contaminated by *Rhizobium japonicum* and that provide for the growth of the nodule

fixation zone, made up of bacteroids, plant cells filled with rhizobia that have taken on a swollen or pleomorphic shape. These rhizobia contain an enzyme, nitrogenase, that enables them to fix atmospheric nitrogen. Nitrogenase contains iron and molybdenum, metals necessary in the transport of electrons required in the reduction reaction. The cells at the fixation site are tinted red by leghaemoglobin, a pigment that transports oxygen to the bacteroid

degeneration zone, green or brown. No fixation takes place in degenerating cells

vascular system derived from vessels of the stele, it irrigates the nodule, bringing in necessary carbohydrates for the fixation reaction and transporting the nitrogen compounds that are formed to the leaves

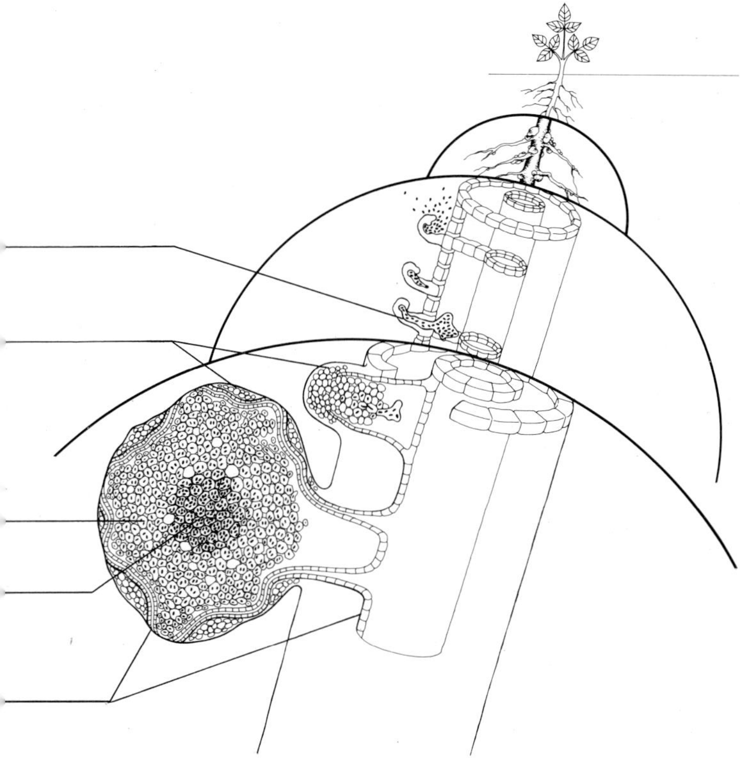

Figure 4. Infection process and nodule development on a soybean (*Glycine max*) root

tiplies, and an infection thread is formed. This thread penetrates the root cortex. Other root cells are infected and cell division increases. The embryonic nodule develops.

Nodules vary widely in shape, size, colour, texture and location. Shape and location of effective nodules are largely determined by the host plant. Nodules produced by a strain of *Rhizobium* on one legume host may have no resemblance to nodules produced by the same strain on another host in the same "cross-inoculation" group.

The size, colour and distribution of nodules on the roots of leguminous plants reflect the status of the *Rhizobium*-legume association and its nitrogen-gathering efficiency. This varies widely with the strain of *Rhizobium*. Various types of effective nodules are shown in Figures 5 and 6.

Nodules that can be seen with the naked eye usually develop in 10 to 14 days on seedlings grown in a nitrogen-free substrate in a greenhouse or growth chamber. The time required varies somewhat, depending on the legume and the seed size. Under field conditions, visible nodules should appear in 21 to 28 days. With high levels of nitrogen in the soil or application of large amounts of nitrogen fertilizer, nodule development will be delayed. In tropical climates, however, small applications of nitrogen fertilizer are often used to stimulate early growth of leguminous crops. The overall effect can be beneficial.

Kinds and distribution of nodules

Effective nodules are generally large and are clustered on the primary and upper lateral roots. Maximum development of nodules as determined by weight or volume normally occurs in the late flowering stage. In contrast, ineffective nodules are small, numerous, and usually distributed throughout the root system. Examples of effective and ineffective nodules on *Trifolium* sp. are shown in Figure 7.

Nodule colour

The effectiveness of the *Rhizobium*:leguminous plant association can also be determined by slicing the nodules of the legume host during the early flowering period and noting the nodule colour. Effective nodules are large and have a deep reddish colour inside. The red pigment, leghaemoglobin, is associated with active nitrogen fixation in legume nodules. It is not a part of nitrogenase, the nitrogen-fixing enzyme, but it controls the oxygen necessary to activate this enzyme.

Effective nodules on the prairie bean (*Strophostyles helvola*), cowpea (*Vigna unguiculata*) and the hyacinth bean (*Dolichos lablab*) are sometimes black on the inside. These nodules usually contain both leghaemoglobin and melanin, the latter causing the black colour.

The red haemoglobin in effective nodules breaks down to green legcholeglobin as the nodules become older and senescent. A single effective nodule on certain legumes — crown vetch (*Coronilla varia*), sainfoin (*Onobrychis viciifolia*) and milk vetch (*Astragalus cicer*) — may show white, red and green regions simultaneously during any stage of growth. These indicate, respectively, areas of nodule growth, active nitrogen fixation and senescence (Fig. 7). Ineffective nodules are white to pale green on the inside. They do not change colour as they age.

When legumes are supplied with nitrogen fertilizer, the nodules produced by effective strains of rhizobia remain small and exhibit the same characteristics as those produced by ineffective rhizobia. After the soil nitrogen is exhausted, the nodules usually increase in size and function normally.

By contrast, nodules produced by effective rhizobia on molybdenum-deficient plants tend to grow large and appear normal except for colour. Molybdenum-deficient nodules are green on the inside and appear senescent.

The pigmentation of nodule interiors is shown in Figure 8.

Figure 5. Kinds of nodules on grain legumes:

(*a*) and (*b*) groundnut (*Arachis hypogaea*)

(*c*) soybean
(*Glycine max*)

(*d*) chick-pea
(*Cicer arietinum*)

Figure 6. Kinds of nodules on forage legumes:

(*a*) birdsfoot (*Lotus corniculatus*)

(*b*) white clover (*Trifolium repens*)

(*c*) sainfoin
(*Onobrychis viciifolia*)

(*d*) guar
(*Cyamopsis tetragonoloba*)

Figure 7. Effective and ineffective nodules:

(*a*) nodules of *Trifolium* spp., ineffective on bottom, effective on top

(*b*) effective nodules of crown vetch (*Coronilla varia*)

Figure 8. Section of soybean (*Glycine max*) nodules. Top to bottom, effective to ineffective

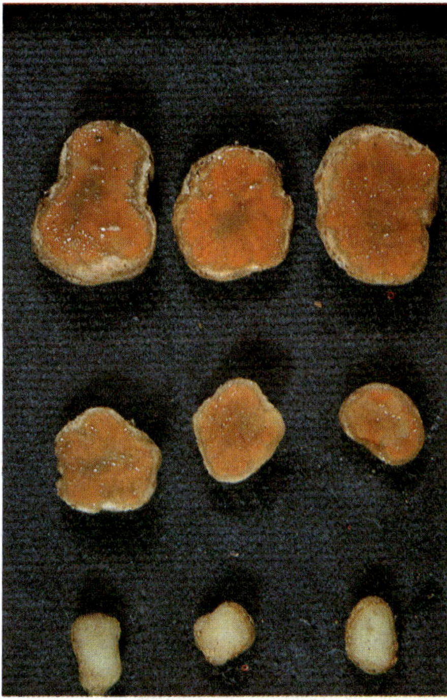

Nodule examination

When nodulation of field plants can be examined only once, the best time is at the peak of flowering. If more frequent examination is possible, nodule development at four to five weeks should reveal adequacy of the inoculation treatment and relative infectiousness of the *Rhizobium* strains. Late examination during pod formation can be useful in selecting the more compatible *Rhizobium* strains during the critical period of seed development. Whenever possible, all three examinations are recommended.

Species and strains of rhizobia: host specificity

Early in the study of the *Rhizobium*:legume association, it was noted that certain plants showed preferences for certain rhizobia and vice versa. Leguminous species mutually susceptible to nodulation by a particular kind of rhizobia constitute a "cross-inoculation" group. Strains of rhizobia capable of nodulating the plants in one of these groups were considered a species regardless of whether nitrogen fixation occurred.

The six *Rhizobium* species established under this system and the corresponding genera of plants that each nodulates are given in Table 2, Part 1.

The "cross-inoculation" concept for grouping legumes and establishing *Rhizobium* species was questioned after numerous inharmonious bacteria-plant responses were discovered. The idea was almost abandoned. No new *Rhizobium* species have been introduced during the past half century. Nonetheless, this grouping of plants and designation of *Rhizobium* species on the basis of nodulation have been useful both to scientists and to manufacturers of legume inoculants. While nodulation does not guarantee any benefit in terms of yield, it results in a beneficial association when soil nitrogen is low.

A system of classifying *Rhizobium* species based on numerical taxonomy has been developed and appears in the ninth edition of Bergey's *Manual of determinative bacteriology*. This is included in Part 2 of Table 2 for clarification during the forthcoming "change-over" period.

The "cross-inoculation" grouping of leguminous plants and designation of *Rhizobium* species is based entirely on nodulation and does not relate to host benefit. Numerous data show that nodules produced by some strains of rhizobia are of no benefit to the growth of their host. These ineffective nodules could even be considered parasitic because the plant provides energy for their growth without receiving any nitrogen in exchange.

Many legumes are "host-specific". While they may be

Table 2. *RHIZOBIUM* SPECIES

Bacteria	Genera of host plants
	Part 1
R. meliloti	*Medicago, Melilotus* and *Trigonella*
R. trifolii	*Trifolium* spp.
R. leguminosarum	*Pisum, Vicia, Lathyrus, Lens*
R. phaseoli	*Phaseolus vulgaris, P. multifloris*
R. lupini	*Lupinus* and *Ornithopus*
R. japonicum	*Glycine max*
Cowpea (type)	*Vigna*, many other genera and species
	Part 2
Fast growers	
R. meliloti	*Medicago, Melilotus*, and *Trigonella*
R. leguminosarum	
biovar.—*trifolii*	*Trifolium* spp.
biovar.—*phaseoli*	*Phaseolus vulgaris, P. multifloris*
biovar.—*viceae*	*Pisum, Lathyrus, Lens, Vicia*
R. loti	*Lupinus, Lotus, Anthyllis, Ornithopus*
Slow growers	
Bradyrhizobium japonicum	*Glycine max*
Bradyrhizobium spp.	
Bradyrhizobium sp. *(Vigna)*	*Vigna* — numerous other genera and species
Bradyrhizobium sp. *(Lupinus)*	*Lotus pedunculatus, Lupinus* sp.

Source: Bergey, *Manual of determinative bacteriology*, 9th ed., 1983.

nodulated by diverse strains of rhizobia, growth is enhanced only when the nodules are produced by particular strains of rhizobia. It is thus extremely important to match symbionts prudently for maximum nitrogen fixation.

Effectiveness grouping

As the study of the responses of numerous legume species to different strains of rhizobia has been expanded, it has been noted that certain plants tend to respond similarly to particular strains of rhizobia. In other words, a strain of rhizobia that nodulates and fixes a large amount of nitrogen in association with one species may also do the same in association with certain other species. This must be verified by testing. Leguminous plants that demonstrate this tendency to respond similarly to particular strains of rhizobia are considered an "effectiveness" group.

The effectiveness groupings of leguminous species are given in Table 3. One inoculant can be prepared for an entire effectiveness group of plants rather than for each individual plant species.

Table 3. EFFECTIVENESS GROUPINGS OF LEGUME CROPS

Legumes that tend to respond similarly when inoculated with the same strain of rhizobia

Rhizobium species	Effectiveness groupings	Leguminous species
Rhizobium meliloti		
	1.	*Medicago sativa, M. falcata, M. minima, M. tribuloides, Melilotus denticulata, M. alba, M. officinalis, M. indica*
	2.	*Medicago arabica, M. hispida, M. lupulina, M. orbicularis, M. praecox, M. truncatula, M. scutellata, M. polymorpha, M. rotata, M. rigidula, Trigonella foenum-graecum*
	3.	*Medicago laciniata*
	4.	*Medicago rugosa*

Table 3. *(cont'd)*

Rhizobium species	Effectiveness groupings	Leguminous species
Rhizobium trifolii		
	5.	*Trifolium incarnatum, T. subterraneum, T. alexandrinum, T. hirtum, T. arvense, T. angustifolium*
	6.	*Trifolium pratense, T. repens, T. hybridum, T. fragiferum, T. procumbens, T. nigrescens, T. glomeratum*
	7.	*Trifolium vesiculosum, T. berytheum, T. bocconei, T. boissieri, T. compactum, T. leucanthum, T. mutabile, T. vernum, T. physodes, T. dasyurum*
	8.	*Trifolium ruepellianum, T. tembense, T. usambarense, T. steudneri, T. burchellianum* var. *burchellianum, T. burchellianum* var. *johnstonii, T. africanum, T. pseudostriatum*
	9.	*Trifolium semipilosum* var. *kilimanjaricum, T. masaiense, T. cheranganiense, T. ruepellianum* var. *lanceolatum*
	10.	*T. medium, T. sarosience, T. alpestre*
	11.	*T. ambiguum*
	12.	*T. heldreichianum*
	13.	*T. masaiense*
	14.	*T. reflexum*
	15.	*T. rubens*
	16.	*T. semipilosum*
Rhizobium leguminosarum		
	17.	*Pisum sativum, Vicia villosa, V. hirsuta, V. faba, V. tenuifolia, V. tetrasperma, Lens esculenta, Lathyrus aphaca, L. cicera, L. hirsutus, L. odoratus, L. sylvestris*
	18.	*Lathyrus ochrus, L. tuberosus, L. szenitzii*
	19.	*Lathyrus sativus, L. clymenum, L. tingitanus*
	20.	*Vicia faba, V. narbonensis*
	21.	*Vicia sativa, V. amphicarpa*

Table 3. *(cont'd)*

Rhizobium species	Effectiveness groupings	Leguminous species
Rhizobium phaseoli		
	22.	*Phaseolus vulgaris, P. coccineus, P. angustifolius*
Rhizobium lupini		
	23.	*Lupinus albicaulis, L. albifrons, L. albus, L. angustifolius, L. arboreus, L. argenteus, L. benthamii, L. formosus, L. luteus, L. micranthus, L. perennis, L. sericeus, Lotus uliginosus, L. americanus, L. pedunculatus, L. strictus, L. strigosus*
	24.	*Lupinus densiflorus, L. vallicola*
	25.	*L. nanus*
	26.	*L. polyphyllus*
	27.	*L. subcarnosus*
	28.	*L. succulentus*
Rhizobium japonicum		
	29.	*Glycine max*
Rhizobium spp. (cowpea type)		
	30.	*Vigna unguiculata, V. sesquipedalis, V. luteola, V. cylindrica, V. angularis, V. radiata, V. mungo, Desmodium* sp., *Alysicarpus vaginalis, Crotalaria* sp., *Macroptilium lathyroides, M. atropurpureum, Psophocarpus* sp., *Lespedeza striata, L. stipulacea, Indigofera* sp., *Cajanus cajan, Cicer arietinum*
	31.	*Phaseolus limensis, P. lunatus, P. aconitifolius, Canavalia ensiformis, C. lineata*
	32.	*Arachis hypogaea, A. glabrata, Cyamopsis tetragonoloba, Lespedeza sericea, L. japonica, L. bicolor*
	33.	*Centrosema pubescens, Galactia* sp.
	34.	*Lotononis bainesii*

Table 3. *(concluded)*

Rhizobium species	Effectiveness groupings	Leguminous species
	35.	*Lotononis angolensis*
Rhizobium spp. (lotus)		
	36.	*Lotus corniculatus, L. tenuis, L. angustissimus, L. tetragonolobus, L. caucasicus, L. crassifolius, L. creticus, L. edulis, L. frondosus, L. subpinnatus, L. weilleri, Dorycnium hirsutum, D. rectum, D. suffruticosum, Anthyllis vulneraria, A. lotoides*
	37.	*Lotus uliginosus, L. americanus, L. scoparius, L. angustissimus, L. pedunculatus, L. strictus, L. strigosus, Ornithopus sativus, Lupinus angustifolius, L. albus, L. luteus*
Rhizobium spp. (*Coronilla, Petalostemon-Onobrychis*)		
	38.	*Coronilla varia, Onobrychis viciifolia, Petalostemon purpureum, P. candidum, P. microphyllum, P. multiflorus, P. villosum, Leucaena leucocephala, L. retusa*
Rhizobium spp. (various)		
	39.	*Dalea alopecuroides*
	40.	*Strophostyles helvola*
	41.	*Robinia pseudoacacia, R. hispida*
	42.	*Amorpha canescens*
	43.	*Caragana arborescens, C. frutescens*
	44.	*Oxytropis sericea*
Rhizobium spp. *Astragalus* sp.		
	45.	*A. cicer, A. falcatus, A. canadensis, A. mexicanus, A. orbiculatus*

4. INOCULANTS AND INOCULATION

In many soils, the nodule bacteria are not adequate in either number or quality. Under these conditions, it is necessary to inoculate the seed or soil with highly effective rhizobia cultures.

Nodule bacteria (*Rhizobium* spp.) are cultured in the laboratory and combined with a suitable carrier material, such as peat, compost or filter mud, to make an inoculant. The process of adding this inoculant to seed or soil is called inoculation.

Qualities of an effective inoculant

(*a*) The inoculant should contain only rhizobia capable of producing nodules and fixing a large amount of nitrogen on the designated hosts. Effective inoculants may consist of a single strain of *Rhizobium* or they may contain several strains. The first type is called a unistrain and the latter a multistrain inoculant. A unistrain inoculant is good where field tests have shown that a particular strain of rhizobia works best on a particular legume host or genotype under the prevailing soil and climatic conditions and where farmers generally plant that particular variety or genotype of legume. A multistrain inoculant is preferable in areas where many varieties or cultivars of legumes may be grown and where there may be wide variation in soil and climate. The multistrain inoculant should not contain any strains of rhizobia that form nodules without benefit to their host or hosts. Such strains can impede nodulation by effective rhizobia.

(*b*) The inoculant should provide large numbers of viable rhizobia, at least 10 000 to a million per seed.

(*c*) The carrier medium or inoculant base must protect the

rhizobia in the package and on the seed. It should be easy to apply and should adhere well to the seed.

(*d*) The inoculum must be free of other bacteria which might be detrimental to rhizobia or to the young legume seedling.

(*e*) The inoculant must be packaged to protect the rhizobia until used by the farmer. The package should allow exchange of gases and should retain moisture.

(*f*) The package should provide clear instructions and list the leguminous species that it nodulates effectively.

(*g*) The package should show the date beyond which the product cannot be considered dependable.

(*h*) The name and address of the manufacturer should be printed on the package.

Types of inoculants

Legume inoculants are of two general types: those designed for application to seeds and those designed for application directly to soil. Seed inoculants are the most common because they are easy to apply and are generally effective under normal conditions. Application of inoculant directly to the soil may be necessary to obtain effective nodulation when planting legume seeds in hot, dry or highly acid soils or under adverse weather conditions, when seeds are treated with chemicals toxic to rhizobia or when the soil harbours a large population of highly infective non-nitrogen-fixing rhizobia.

High-quality peat-based inoculants are generally considered the most dependable. However, the shelf life or period of safe use will vary with peat quality and treatment as well as storage temperature. It is important that each country study and formulate a dating system for inoculant that is realistic and protects the interests of the farmer or grower.

Two types of peat-based inoculants, one for seed and one for soil, are shown in Figure 9.

Figure 9. Seed and soil variety peat-based inoculants: (*a*) commercial packaged inoculants; (*b*) powder inoculant on left, granular inoculant on right

Quality control and standards

Viable rhizobia may be expressed in terms of number per gram of inoculant or as number per seed. The latter is more meaningful because the quantity of inoculant per unit weight of seed varies with different inoculant manufacturers.

In Australia, inoculants must contain a minimum of 1×10^9 viable rhizobia per gram at manufacture and 1×10^8 per gram up to the expiration date. In Canada and New Zealand, the standards for peat inoculants are similar. Canadian authorities have adopted the following standards for pre-inoculated seeds: small seeds (e.g., clover), 1 000/seed; intermediate-sized seeds (e.g., sainfoin), 10 000/seed and large seeds (e.g., soybean), 100 000/seed.

In France, farmers' practices are used in establishing standards. Inoculants must provide a minimum of 5×10^3 viable rhizobia per lucerne seed and 1×10^6 viable rhizobia per soybean seed up to the expiration date of the inoculant. It is assumed that the inoculant will be stored at about 20°C for as long as six months before using. With these constraints, the manufacturer needs to ensure that inoculants contain at least 10^8 or 10^9 viable rhizobia per gram when produced. This approach is recommended for all countries after rhizobia survival under local conditions has been determined for the various brands of inoculant marketed.

It is now recognized that large numbers of viable effective rhizobia are required to inoculate legumes successfully. The old adage "safety in numbers" is particularly apt when inoculating legume seeds.

Tests for viable rhizobia

Tests for numbers of viable rhizobia in inoculants are made by one of two methods: dilution plates using differential media, or nodulation tests in which growing plants are inoculated with a series of dilutions of the inoculum and

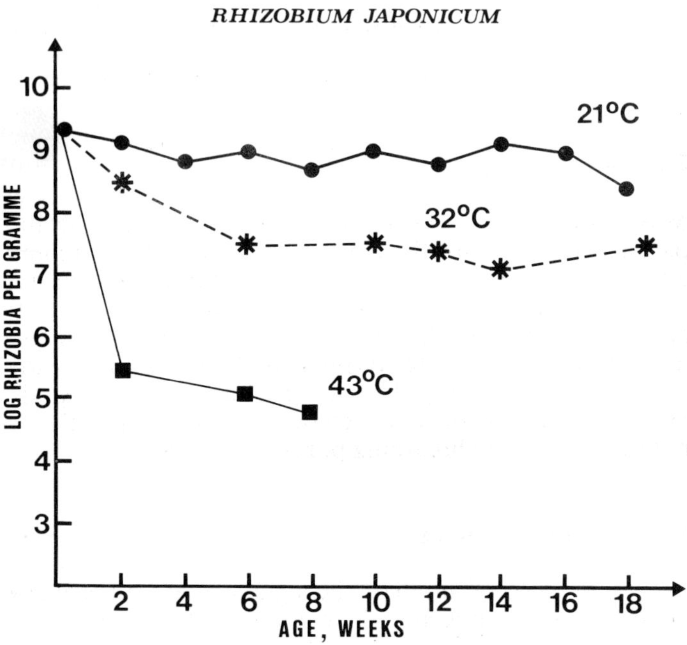

Figure 10. Survival of *Rhizobium japonicum* in a peat-base inoculant packed in polyethylene as affected by different storage temperatures

reference is made to most probable number (MPN) tables. Each method has advantages and disadvantages. Both have a common fault in that they require two to four weeks to make a determination. Quality may be lost during the period required to conduct the test.

Rhizobia are living one-celled plants and will die if the temperature is too high or if there is insufficient moisture. Survival of *Rhizobium japonicum* at three different temperatures in a peat-base inoculant is shown in Figure 10.

Unfortunately, an inoculant containing only dead rhizobia looks just like one with live rhizobia. Furthermore, some live rhizobia are unable to produce nodules. It is therefore necessary to test not only for number of live rhizobia but for ability to produce effective nodules on the specific host. This requires considerable time.

Packaging

Most manufacturers of legume inoculants use a flexible film package such as polyethylene, which retains moisture and allows the desired exchange of gases. Films that do not allow diffusion of oxygen and carbon dioxide are not suitable. The packaging film may vary from 0.027 to 0.076 millimetres (0.003 to 0.005 inches) in thickness to provide adequate strength to withstand the hazards of shipping. High-density polyethylene may be used if contamination is not of concern and breathing pores are provided.

Labelling requirements

The information required on the legume inoculant package varies by country but should include the following:

(*a*) Name or names of the leguminous plants for which the inoculant is considered effective (common and scientific names)
(*b*) Scientific name of the *Rhizobium* species
(*c*) Number of live rhizobia per gram or number of viable rhizobia per seed that the product will provide when used according to directions
(*d*) Expiration date — date beyond which the product cannot be considered dependable and should not be used
(*e*) Lot number for quality control
(*f*) Instructions for use
(*g*) Net weight of inoculant
(*h*) Trade name, manufacturer and address
(*i*) Storage conditions required (remember it is a perishable product)

Shelf life and dating system

Legume inoculants are perishable and can lose their effectiveness when exposed to a temperature of 40°C or higher for a few hours. High-quality inoculants should retain their effectiveness for six months or longer when stored at a temperature around 20°C. This period can be extended if inocula are refrigerated near 4°C. Warning should be given regarding the harmful effects of high temperatures on inoculants. Warning should also be given about the harmful effects of deep-freezing. Deep-freezing of inoculants may damage the *Rhizobium* cell membranes.

Hints for buying and storing inoculants

(*a*) Different legumes require different inoculants. Be certain to have the right one. The legume name should be given on the package.

(*b*) Be certain the inoculant is fresh. Look for the expiration date beyond which the inoculant cannot be depended upon to produce good results. Use high-quality inoculant.

(*c*) Keep the inoculant in a cool place until ready to use it. Storage in a refrigerator is good. Cold will not harm the bacteria but they cannot tolerate high temperatures.

(*d*) It is best to inoculate seeds just before planting. Rhizobia die quickly on the seed. Use high-quality seed. Plant in a well-prepared seed-bed.

(*e*) When inoculating seeds, avoid making them too wet. Use just enough water for good adhesion of the inoculant to the seed.

(*f*) Chemicals on seeds, and fungicides or insecticides applied to the soil may be toxic to the nodule bacteria. If seeds are chemically treated, use a larger amount of inoculant and plant immediately after inoculating.

(*g*) Store inoculated seed in a cool protected place until planting. Keep it out of hot sunshine and protect from excessive drying.

(*h*) Leftover inoculant may be kept safely in the package provided that it is closed tightly to prevent excessive drying. Leftover inoculant stored in a refrigerator at 4°C or lower will remain effective for several months.

(*i*) Nodules should develop on legume roots in three or four weeks. When examining for nodules, use a shovel or fork for digging the plants. The nodules may be stripped from the roots if the plants are pulled.

(*j*) Leftover inoculated seed should not be eaten or fed to animals.

5. DETERMINING THE NEED FOR INOCULATION

Inoculation is almost always needed when certain new leguminous crops are introduced to new areas or regions. It is often beneficial to inoculate newly developed or introduced cultivars of legumes even though the same species may have been grown previously. Host-specific rhizobia are frequently developed for new cultivars or varieties of legumes.

Many soils are heavily infested with ineffective rhizobia capable of inducing nodulation without host benefit. Under such conditions, a very large inoculum of competitive and highly effective strains of rhizobia is needed to counteract the aggressive native rhizobia.

Field test

When the situation is not known, a field test to determine the need for inoculation is recommended. Three basic treatments are needed:

(*a*) Inoculated plants from seeds inoculated with the best inoculant available
(*b*) Non-inoculated plants receiving no fertilizer treatment
(*c*) Non-inoculated plants furnished with fertilizer nitrogen (to serve as a test for growth conditions).

Whenever possible, these three treatments should be made at two fertility levels: that which would be used by the farmer and the optimum level from an economic viewpoint. Four replications of each treatment are recommended and plots should be randomized. A model experimental design for a legume inoculation trial is presented in Figure 11.

Reactions and interpretation

An explanation of different situations which might develop in these field trials is given in Table 4.

6. METHODS OF INOCULATION

To form an inoculant, laboratory-grown cultures of rhizobia are mixed with various inert, finely pulverized solids, such as peat, compost, filter mud, bagasse or some other suitable carrier. This formulation is usually applied to legume seeds before planting to make certain that effective nodules will develop and that the plants will have a dependable supply of nitrogen. An alternative method is to apply the inoculant directly to the soil. This is necessary under some conditions although it may be more expensive.

With seed, the inoculant rate recommended by manufacturers ranges from 4 to 6 grams per kilogram of seed or

TREATMENTS:

TREATMENT*	NITROGEN	INOCULATION
F1	−	−
F2	+	−
F3	−	+
M1	−	−
M2	+	−
M3	−	+

*F INDICATES FARM FERTILITY; M INDICATES MAXIMUM FERTILITY

PLOT LAYOUT:

REP 3													REP 4
NUMBER	13	14	15	16	17	18	19	20	21	22	23	24	
TREATMENT	F3	M1	M2	F1	M3	F2	M2	M1	F3	F2	M3	F1	
REP 1													REP 2
NUMBER	1	2	3	4	5	6	7	8	9	10	11	12	
TREATMENT	F1	F3	F2	M2	M3	M1	F1	M3	M1	F2	M2	F3	

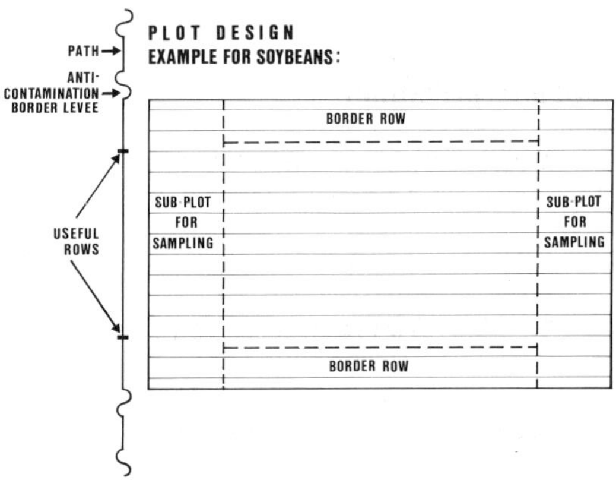

Figure 11. Model experimental design for legume inoculation trial

0.28 to 0.42 kilograms per hectare at the usual seeding rate of cowpeas (*Vigna unguiculata*) or soybeans (*Glycine max*). By comparison, 6 to 8 kilograms of soil granular inoculant are required per hectare of land.

Table 4. EXPLANATIONS FOR FIELD SITUATIONS FOUND IN INOCULATION TRIALS

Situation	Explanation
Non-inoculated plants	
1. No nodules on non-inoculated control. Plants yellow.	No native rhizobia capable of infecting that legume.
2. Many small nodules scattered over root system. Plants yellow.	Native rhizobia are ineffective in fixing nitrogen.
3. No nodules on non-inoculated control. Plants green.	Soil high in mineral nitrogen. No native rhizobia capable of nodulating that legume.
4. Small nodules on non-inoculated control. Plants deep green.	Soil high in mineral nitrogen. Native rhizobia may be effective or ineffective.
5. Non-inoculated control plants have numerous large nodules. Plants deep green colour.	Native rhizobia infective and effective. Inoculation not necessary.
Inoculated plants	
6. Nitrogen control — nodulated — nodules small, plants green.	Infective rhizobia. Nodules not active because of nitrogen level.
7. Inoculated plants have no nodules. Plants yellow.	Improper inoculum or rhizobia in inoculum dead.
8. Inoculated plants have small nodules and deep green colour.	Soil high in mineral nitrogen. Nodules not operating.
9. Inoculated plants have large nodules, red on the inside. Plants deep green. Non-inoculated control has small nodules or no nodules and yellow colour.	Native rhizobia ineffective. Inoculant highly effective.
10. Inoculated + nitrogen — plants larger and greener than those receiving only inoculant. Nodules small to medium.	Rhizobia not adequate. Need more effective strains.
11. Inoculated plants receiving phosphorus and potassium — larger, more vigorous than non-fertilized plants from inoculated seed.	Soil low in phosphorus and potassium. Needs fertilizer for maximum nitrogen fixation.

Seed inoculation

Forms of seed inoculants

The forms of seed inoculants currently available are:

(*a*) Agar slants in bottles
(*b*) Peat, compost, coal, vermiculite, and other solid-based moist powders
(*c*) Broth or liquid cultures
(*d*) Lyophilized or freeze-dried preparations
(*e*) Frozen concentrates of broth
(*f*) Oil-dried preparations on talc or vermiculite
(*g*) Rhizobia entrapped in polyacrylamides, alginates or xanthanes

Methods of applying seed inoculants

Slurry. In the slurry method (Fig. 12), the inoculant is first mixed with water to form a uniform, pourable suspension. In some instances, gums or sugar may be added to the water to improve adhesion of the inoculant to the seed. The use of sugar in the inoculant slurry decreases the death rate of rhizobia on the drying seed.

The amount of water needed in the slurry varies with different seeds. More water is needed for small seeds than for large seeds because of the greater surface area to be coated. It is very important not to get the seed too wet because seeds may stick together or be damaged by the seed drill. The seeds should be thoroughly coated with the inoculant particles (Fig. 13).

The amounts of water needed to inoculate various kinds of seed are given in Table 5.

Seed hopper — dry method. In this method, the inoculant powder is mixed directly with seed without using any water or other liquid. Because the dry inoculant does not adhere well to the seed and is lost, it is generally unsatisfactory

Figure 12. Seed inoculation by slurry method: (*a*) view of equipment, (*b*) addition of water to peat-base inoculant

Figure 12. (*c*) mixing of slurry with seed, (*d*) seed coated with dried inoculant

Table 5. AMOUNTS OF INOCULANT AND WATER REQUIRED FOR SLURRY INOCULATION OF LEGUME SEEDS OF VARIOUS SIZES

Legume species	Seeds (no./kg) [1]	Inoculant (g/25 kg of seed)	Water (ml/25 kg of seed)	Slurry (ml/25 kg of seed)
Trifolium repens (white clover)	2 000 000	110	625	750
Medicago sativa (alfalfa)	500 000	110	550	650
Coronilla varia (crown vetch)	250 000	110	550	650
Vigna radiata (green gram)	25 000	110	500	550
Vigna unguiculata (cowpea)	10 000	110	375	437
Glycine max (soybean)	5 000	110	250	287
Cicer arietinum (chick-pea)	2 000	110	250	287
Vicia faba (broad bean)	1 250	110	175	200

Note: Legumes are arranged in ascending order of seed size.
[1] Approximate values.

Figure 13. Soybean (*Glycine max*) seeds: slurry inoculated on top, non-inoculated on bottom

and is not recommended. Its only advantage is that it is easy.

Lime pellet. Rhizobia die on inoculated seed when the seeds are planted in highly acid soil or when the seeds are mixed with acid fertilizers prior to sowing. In order to avoid this, legume seeds are lime-pelleted. First, the seeds are inoculated by wetting them thoroughly with a peat-base inoculant slurry. Then they are quickly mixed and rolled in a finely pulverized limestone until they are thoroughly

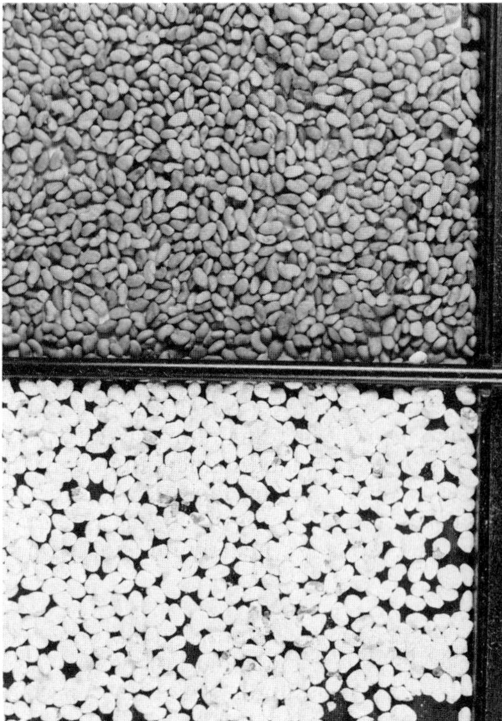

Figure 14. Lucerne (*Medicago sativa*) seeds: non-inoculated on top, lime coated on bottom

coated with the limestone particles (Fig. 14). Under tropical conditions, powdered, rock phosphate is often substituted for the limestone because limestone can be harmful to acid-tolerant rhizobia. The particulate matter limestone and/or pulverized rock phosphate serves to dry the seeds and keep them from adhering to each other.

The inoculant slurry should be made with an adhesive, either gum arabic or a synthetic cellulosic gum, to bind the lime coating firmly to the seeds. The gum solution for making the slurry should be moderately viscous but flow easily.

The proper amounts of inoculant, water, slurry and powdered limestone required for various legume seeds are given in Table 6.

There are some disadvantages to pelleting: the weight of seed is increased substantially, coated seeds are very abrasive and may damage the drill, and special machinery is needed to mix the wet seed with the powdered limestone or phosphate. Also, it is difficult to coat seeds uniformly. The seed/limestone ratio may vary widely and result in irregular seeding rates.

Inoculant pellet. A modified pelleting system in which the seed is coated with peat-base inoculant can be used when a very large *Rhizobium* inoculum is required. One third of the inoculum (4 g/kg seed) is applied to the seed as a gum arabic slurry. The remaining two thirds of the inoculum (8.5 g/kg seed) are mixed immediately with the wet seed. This serves to coat and dry the seed. This system of inoculating has proved very successful for seeds planted in hot, dry soils or where soils are heavily infested with native ineffective rhizobia.

Coated seed. In pelleting legume seeds with pulverized limestone, rock phosphate or peat-base inoculant, water-soluble gums are used. With these vegetable gums, the pelleting materials tend to flake off, particularly when the seed is inoculated several weeks in advance. To overcome this, various water-insoluble adhesives such as polyvinyl acetate, polyvinyl alcohol, polyurethane, polyurea varnishes or resins are dissolved in solvents and used to bond the particulate substances — charcoal, clay, diatomaceous earth, talc and other substances — to the seed. Peat-base inocula, seed chemicals and micronutrients may also be included in the mix. The seed coating is usually done by seed processors before the seeds are offered for sale. Usually it is not practicable to do the seed coating on the farm.

Some of the advantages claimed for coated seed are: more uniform seeding, easier planting, better germination,

Table 6. AMOUNTS OF INOCULANT, WATER, SLURRY, AND POWDERED LIMESTONE REQUIRED TO PELLET LEGUME SEEDS OF VARIOUS SIZES

Legume species	Seeds (no./kg)	Water (ml/25 kg of seed)	Slurry (ml/25 kg of seed)	Limestone (ml/25 kg of seed)
Trifolium repens (white clover)	2 000 000	1 050	1 175	10
Medicago sativa (alfalfa)	500 000	1 050	1 175	10
Coronilla varia (crown vetch)	250 000	1 050	1 175	10
Vigna radiata (green gram)	25 000	950	1 000	8.7
Vigna unguiculata (cowpea)	10 000	425	500	5
Glycine max (soybean)	5 000	425	500	5
Cicer arietinum (chick-pea)	2 000	400	475	5
Vicia faba (broad bean)	1 250	375	425	5

Note: The inoculum is used at a rate of 250 grams per 25 kilograms seed, which is approximately twice the normal rate.

stronger and healthier seedlings, early nodulation, better nourished seedlings and so on. However, the beneficial effects of these coatings have not been proved. Further, solvent-based adhesives are often toxic to the rhizobia. With a very large peat-base inoculum, some rhizobia may survive for a few weeks, but prospects for effective nodulation under most field conditions are poor.

Pre-inoculated seed. Pre-inoculated seed is seed inoculated before being offered for sale. The inoculation is usually done by seed processing companies months in advance of planting. While this idea is attractive to farmers, results have been very disappointing. The legume seed does not provide a good habitat for rhizobia; the death rate is rapid. A peat-base inoculum applied as a coating with gums and sugars has assured better survival than other systems, but results with pre-inoculated seed have not been dependable.

Soil inoculation

Necessity

Soil inoculants are those designed for application directly to the soil rather than to the seed (Fig. 15). This type of inoculation is recommended under the following conditions:

(*a*) When the leguminous seeds are coated with toxic chemicals for protection against fungi, insects or other soil pests and rhizobia applied to seeds would be killed.
(*b*) When planting in hot, dry soils is unavoidable. Placing rhizobia in moist soil below the seeds can result in good survival and effective nodulation.
(*c*) When soils are heavily infested with large populations of infective non-nitrogen-fixing strains. Soil inoculants can introduce a larger inoculum of effective rhizobia than can be applied directly to the seed.

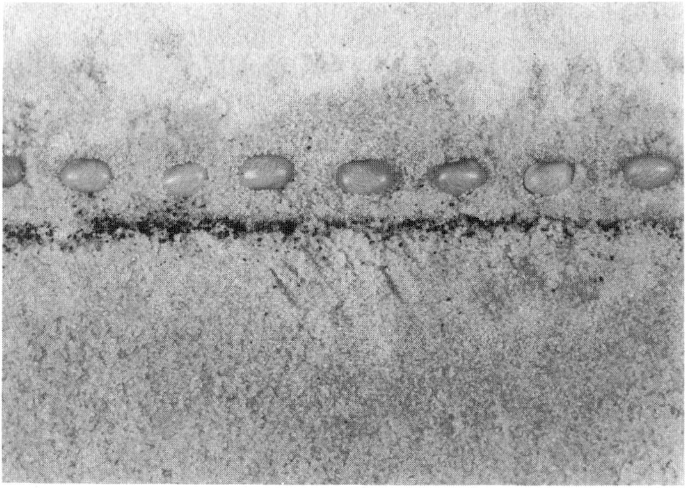

Figure 15. Soil inoculation. Side view showing layer of inoculant granules placed below the seed (bean, *Phaseolus vulgaris*)

(*d*) When good stands of legumes are obtained initially but for some reason the plants fail to develop nodules. Soil inoculants can be drilled into the ground after planting using sod-type seeders, or broadcast or sprayed on the surface soil just before rain or irrigation. This emergency inoculation practice can bring about effective nodulation and save the stand.

Forms of soil inoculants

The forms of soil inoculants vary widely, depending upon the preferred method of application. The various types of soil inoculant are:

Liquid or frozen concentrates. In the Netherlands, these have been distributed successfully for many years using drip or spray systems. Effective inoculation of soybeans was obtained in the Casamance area of Senegal by applying 5 litres of a slurry (2/3 peat and 1/3 water) per hectare. In Australia, good results are obtained by mixing the regular peat-based seed inoculant with water and pumping the resulting suspension into the seed furrow using a drill-mounted pumping unit.

Inoculated granule. Small marble, calcite or silica grains or cores are wetted with a good adhesive and inoculated with a powder-type (usually peat-base) inoculum. After drying, the granules are distributed by broadcasting, using a whirlwind-type spreader. The dense granules are very suitable for gravity distribution either on the surface or from the air.

Granular inoculants are prepared in New Zealand by binding a finely pulverized peat inoculant to coarse sand using an adhesive. This inoculant is broadcast over the soil surface or applied with a drill attachment for granular products.

Granular soil inoculants designed for distribution through drill applicators built for insecticide application are of more recent origin. These enable uniform distribution of a relatively small amount of inoculant, 6 to 8 kilograms per hectare.

Natural peat granule. This form of inoculant "Implant" is prepared by adding *Rhizobium* broth culture to natural sedge peat granules ranging from 300 to 800 μm in diameter. Sufficient moisture (32 to 35 percent) is added to permit growth and multiplication of the rhizobia in the granules. The inoculant flows easily and can be distributed uniformly using a granular applicator drill attachment.

Porous gypsum granule. This granule is prepared by extruding calcium sulphate paste through small circular openings and segmenting with a knife to obtain the desired length. The gypsum granules are made in different sizes to correspond with seed size. After drying, the gypsum granules are sprayed with a broth culture of rhizobia mixed with 12.5 percent milk powder and an equal volume of saturated sucrose solution. The gypsum granules are then ready to be mixed and distributed simultaneously with the seed.

Other granules. In France, three inoculants — a polyacrylamide (PER), an alginate (AER) and an xanthan/carob gum (XER) inoculant — have been used successfully in inoculating soybeans grown in soil free of *Rhizobium japonicum*. In trials, the semi-dried XER form produced the best results. Application rates varied from 26 to 54 kilograms per hectare; only the high rate produced a significant increase in yield over the seed-applied inoculant.

7. COMPATIBILITY OF RHIZOBIA WITH PESTICIDES AND MICRONUTRIENTS

Legume seeds are often treated with chemical insecticides and fungicides to protect the seed and seedlings from harmful insects and soil microorganisms during germination and early seedling development. In some cases, micronutrients such as molybdenum, cobalt or iron, which are required only in minute amounts, may also be applied to seed before sowing for convenience.

This is an inexpensive method to obtain uniform distribution. When chemically treated seeds are inoculated, the rhizobia may be adversely affected by any one or a combination of seed dressings. Since large numbers of viable rhizobia are required to bring about effective nodulation, compatibility or survival of rhizobia on the treated seeds is a major concern.

The literature on compatibility of rhizobia with chemical seed dressings is controversial because tests are often made using soil in jars or from fields that harbour sufficient native rhizobia to bring about good nodulation. Chemical dressings applied to seed will have little effect on the rhizobia already in the soil. Experiments to study compatibility should be carried out in soils which are free of rhizobia capable of nodulating the test plant.

Effects of fungicides

Chemical dressings consisting of any form of heavy metal, such as mercury, zinc, copper or lead, are toxic to rhizobia and not compatible with inoculation. Such seeds should not be inoculated directly; soil inoculants must be applied.

Organic fungicides are generally less toxic than the heavy metals but most of these are also toxic. In some cases, the chemical kills the bacteria. In others, the rhizobia may remain viable and produce colonies but lose their ability to induce nodulation. Measurements of compatibility should include not only ability to produce colonies but also ability to induce nodulation and to fix nitrogen.

The following organic chemicals employed as fungicidal treatments for leguminous seeds are generally toxic to rhizobia:

Captan	N-trichloromethylthio-4-cyclohexene-1,2-dicarboximide
Carboxin	5,6 dihydro-2-methyl-N-phenyl-1,4 oxathiin-3-carboxamide
Chloranil	2,3,5,6 tetrachloro-1,4 benzoquinone
PCNB	pentachloronitrobenzene
Thiabendazole	"Tecto" 2-(4' thiazolyl)-benzimidazole
Thiram	tetramethyl-thiuram-disulphide, ($[(CH_3)_2NCS]_2S_2$)

Thiram appears to be an exception, but under some conditions can also be harmful, particularly with long exposure.

Insecticides

Compatibility of rhizobia with insecticides and herbicides has not been studied extensively. The insecticides most commonly used in legume seeding are Carbofuran 2,3-dihydro-2,2-dimethyl-benzofuranyl methyl-carbamate, Phorate 0-0-diethyl S-(ethylthiol)-methyl phosphorodithioate and Aldicarb 2-methyl-(methylthio) propionaldehyde-o-(methyl-carbomoyl) oxime. They are usually formulated in granules for distribution in the furrow with the seed but are not applied to the seed directly. When used in this manner at recommended rates, these insecticides have had no adverse effect on nodulation when the inoculant was applied to the seed or soil.

Micronutrients

With large legume seeds such as peas or soybeans, it is a common practice to add the needed molybdenum (14 to 20 g Mo/ha) as ammonium or sodium molybdate in the water used in the inoculant slurry. Provided that the seeds are inoculated and planted soon afterwards the rhizobia are unharmed because the salt concentration is relatively low.

With small seeds such as alfalfa or clover that are planted at the low seeding rates of 10 to 12 kilograms per hectare, molybdenum cannot be added to the inoculant slurry because the salt concentration would need to be high (25-28 percent) in order to provide the required amount of molybdenum per hectare. Most of the rhizobia would be killed by osmosis within a few minutes.

The requirement for cobalt is much lower than that for mobybdenum and it can be added to inoculant slurry for large seeds but not for small seeds.

8. MINERAL NUTRITION OF THE SYMBIOSIS

A well-nodulated leguminous plant has its own nitrogen supply that is not immediately available to weeds or companion crops which might be competing for moisture, mineral nutrients or space. The nodulated legume thus has a big advantage, particularly in a nitrogen-poor soil. Nonetheless, for optimum growth and nitrogen fixation it depends upon adequate supplies of all the other essential elements: phosphorus, potassium, calcium, magnesium, sulphur, iron, manganese, boron, zinc, copper, cobalt, molybdenum and chloride. Rhizobia working in nodules provide only the nitrogen.

As a family, the legumes have been misreported as nutritionally demanding and highly intolerant of soil acidity. However, with the exception of alfalfa, *Medicago sativa*, which requires calcium and pH values of 5.5 to 6.0 for one to three days after inoculation for infection to occur, these claims have not been justified. Nearly all of the tropical forage legumes are adapted to acid soils; *Neonotomia wightii* and *Leucaena leucocephala* are exceptions.

There is no good evidence that legumes grown without rhizobia differ in their nutritional needs from other diverse plant groups. However, the need of legumes for adequate mineral nutrition must be recognized.

Acidity, calcium, aluminium, manganese

Acidity and calcium concentration have an interacting effect on rhizobial proliferation and infection of certain legumes. Under field conditions, it is difficult to assess the specific role of each factor on bacterial proliferation and root-hair infection. Calcium serves another role; it moderates the toxic effects of the aluminium and manganese ions on plant growth, which can also limit nitrogen fixation.

The effects of excess aluminium on nodulation and nitrogen fixation apart from pH have not been established. Excess aluminium is known to limit both growth and root elongation. However, high concentrations of aluminium are not obtained until the soil pH is 5.0 or lower; with the exception of cowpeas, most legumes will not nodulate at low pH regardless of the aluminium concentration.

Phosphorus, sulphur and potassium

Deficiencies in phosphorus, sulphur and potassium are most likely to be manifested in decreased growth of the legume, which in turn reduces total nitrogen fixation, but nodule formation itself can also be affected. Field studies with soybeans in Virginia, United States, showed that application of either phosphorus or potassium alone increased the number of nodules per plant and per unit volume of soil. Applied potassium increased the number of nodules, total weight of nodules and pods per plant more than applied phosphorus, but increases were larger when both were supplied. These results indicate the importance of balance. Apparently the supply of phosphorus in this soil was more adequate than that of potassium. The opposite could be expected in acid tropical soils high in aluminium and manganese, where phosphorus is more likely to be deficient.

Sulphur deficiency can limit both nodulation and nitrogen fixation but, again, the effects are most likely to be manifested in reduced growth of the host. Since sulphur is a constituent of many amino acids, a deficiency of this element could result in a lower plant protein yield even without limiting growth.

Nitrogen

It is generally accepted that combined nitrogen in the root medium (soil or substrate) delays or inhibits nodulation.

When fixed nitrogen is available, the leguminous plant, even when nodulated, uses it preferentially. When present, the nodules remain more or less inactive but ready to function once the soil nitrogen has been exhausted. Nodules may remain small and inactive during the entire growing season when there is a large reservoir of nitrogen available in the soil.

Because of the adverse effect of combined nitrogen on nodulation and nitrogen fixation, adding fertilizer nitrogen is not usually recommended for leguminous crops. There are exceptions, however, where prudent use of fertilizer nitrogen has increased yields and actually stimulated nitrogen fixation. When application of small amounts of fertilizer nitrogen stimulates growth of the plant and improves its photosynthetic capabilities without excess delay in nodule development, the overall effects can be beneficial. Small applications of fertilizer nitrogen are often used to stimulate early growth of leguminous crops, particularly in tropical climates.

The effects of combined nitrogen depend to some extent on host species and the strain of rhizobia. Legumes with a short maturity cycle and therefore a short period during which nitrogen fixation can occur are most apt to benefit from wise use of fertilizer nitrogen.

When studying the effects of nitrogen, it should be noted that rhizobial strains vary in the time it takes them to produce visual nodules on the roots and to begin symbiotic nitrogen fixation. With slow or late nodulating rhizobia, increased early growth of the host can increase photosynthetic capacity and nitrogen fixation once the nodule is functioning.

Micronutrients

The nitrogen-fixing enzyme nitrogenase is composed of molybdenum and iron. Without adequate quantities of these elements, nitrogen fixation cannot occur. With seeds such as peas or soybeans, the needed molybdenum (20 to 22 g/ha) can be added in the slurry when the seeds are inoculated (see Chapter 7 above). This is not practicable with smaller seeds sown at much lower rates.

Iron is a constituent of the red leghaemoglobin in the nodule that protects nitrogenase from oxygen inactivation. It is not feasible to apply iron to the seed, but chelated iron can be applied as a foliar spray when legumes show iron deficiency. This is most likely to occur in alkaline soils.

Boron is involved in meristematic activity in both the nodule and the host plant. Without boron, the nodules cannot function.

Zinc, manganese, chloride, iron and cobalt are needed for host growth but do not affect nodulation. A copper deficiency results in development of numerous small nodules typical of those associated with a completely ineffective strain, but the specific role of copper is not understood.

9. INOCULATION COSTS

The cost of legume inoculation varies widely, depending on the species of legume, the treatment used and the country where it takes place. Generally, seed inoculation is not expensive. Approximate costs for seed inoculation of various legume species in the United States (1983) are given in Table 7.

The cost of soil inoculation using peat granules is approximately US$18 for one hectare using an application rate of 6.5 kg. The amount of inoculant applied varies with type of legume and row width, but 10 to 12 kilograms per hectare are adequate for most legumes even when sown broadcast.

Where a leguminous crop is being grown for the first time on soil that contains no rhizobia able to produce effective nodulation so that all of the nitrogen fixed is brought

Table 7. COST OF SEED INOCULATION OF VARIOUS LEGUME SPECIES IN THE UNITED STATES (1983)

Legume species	Seeding rate (kg/ha)	Inoculant cost (US $/ha)	Labour (US $/ha)	Total (US $/ha)
Trifolium repens	3	0.12	0.12	0.24
Clitoria ternatea	4	0.48	0.13	0.61
Leucaena leucocephala	5	0.60	0.17	0.77
Medicago sativa	13	0.69	0.43	1.12
Cajanus cajan	11	1.30	0.37	1.67
Glycine max	65	0.77	2.12	2.93
Vigna unguiculata	76	2.88	2.53	5.41
Vicia faba	87	3.56	2.90	6.46

about by the inoculation, the benefit/cost ratio is very high. Using the inoculation costs in Table 7, two examples are given.

With white clover (*Trifolium repens*) fixing 200 kg/ha of nitrogen and the cost of fixed nitrogen in fertilizer at \$0.50/kg:

$$\frac{200 \times \$0.50}{\$0.24} = \text{416:1 benefit/cost ratio for white clover}$$

With soybeans (*Glycine max*) fixing 100 kg/ha of nitrogen and the cost of fixed nitrogen in fertilizer at \$0.50/kg:

$$\frac{100 \times \$0.50}{\$2.93} = \text{17:1 benefit/cost ratio for soybeans}$$

In France, the benefit/cost ratio has been calculated as 7 for lucerne and 37 for soybeans.

Another method of assessing the benefit/cost ratio of legume inoculation is by yield increase. In farm trials in Burma, inoculation has been credited with 20 to 60 percent increases in yield with chick-peas, groundnuts and green gram. Soybean yields have been increased by two to sevenfold in the Philippines as a result of inoculation. The potential benefit from inoculation is great and the cost is low.

In soils which have successfully grown the legume crop previously, it would be necessary to determine the increase in yield attributed to inoculation rather than the total amount of nitrogen fixed. The benefit/cost ratio would naturally be much smaller.

10. EXTENSION WORK ON INOCULATION

The unique symbiosis of rhizobia with leguminous crops is a fascinating story that will interest many farmers. As extension leaders, you have the opportunity to tell farmers about this remarkable opportunity to utilize nodule bacteria to produce greater yields of higher quality food at a lower cost. By teaching farmers about the benefits of inoculation through conversations, meetings, demonstrations and written articles, you can help them enjoy greater profits and higher living standards.

Making plans

The first step is to introduce local farmers to legume inoculants. Find out where you can obtain effective inoculants for leguminous crops adapted to the area and also literature and pictures which may be used as teaching materials. Plan meetings with farmers to discuss the project. Put on a demonstration on seed inoculation. Explain the results that can be expected. If possible, dig up some native leguminous plants with nodules. Slice nodules and show how to identify effective and ineffective ones. If no growing nodulated legumes are available, use pictures.

Conducting a demonstration

Select the legume that is most important for food locally, preferably an introduced legume. Select a farmer who is interested and willing to work with you in conducting the test. He should be a leader and an innovator.

What you will need to conduct the trial:

(*a*) Fresh inoculant of the proper kind for the legume crop
(*b*) Seeds, if the farmer does not have them
(*c*) A bucket or plastic bag to hold the seeds while they are being inoculated
(*d*) A clean, cool and drinkable water supply
(*e*) Spoons or cups for measuring inoculant, water and seeds (Fig. 16)
(*f*) Stakes to mark plot boundaries
(*g*) Notebook for plot records
(*h*) Shovel or digging tool for examining nodules on roots during growing season
(*i*) Harvesting equipment
(*j*) Scale or measuring container for determining yield

Setting up the field experiment

The field selected should be uniform in fertility, level, low in nitrogen and conveniently located. Demonstration plots may be small but should be large enough to provide a convincing demonstration. A plot of about 0.04 hectare (1/10 acre) is suitable. Plots generally should be rectangular and laid out side by side. Paths of 0.5 to 1 meter wide should be

Figure 16. Measures and containers convenient for inoculating small quantities of legume seeds

left between plots. Plot treatments should be randomized; non-inoculated plots should be planted first to reduce chances of contamination.

For row crops, adjust plot width to give the desired number of rows. At least four rows should be included in each plot; the two centre rows should be saved for harvest. The outer rows may be used to study nodulation. In studying nodulation, plants should not be pulled; roots should be dug out carefully and the soil washed off gently so that the nodules will not be lost.

Inoculating the seed

Since only a small amount of seed will be inoculated in each treatment, measuring spoons and cups are very convenient

Table 8. AVERAGE WEIGHTS OF VARIOUS LEGUME SEEDS

Kind of seed	Approximate amount of seed		
	1 cupful (240 ml)	30 ml	1 tablespoonful
	grams		
Lucerne (alfalfa)	200	25	12
White clover	210	26	13
Vetch	200	25	15
Soybean	165	21	14
Cowpea	190	24	16
Horse bean	200	25	17

for this purpose. With such small amounts of seed, it will probably be easier and more accurate to measure rather than to weigh it. The average weights of several measures of various legume seeds are given in Table 8.

A teaspoon holds about 5 millilitres of water. One tablespoonful is equal to three teaspoonfuls. In inoculating small quantities of seed, it is usually easier to prepare an inoculant slurry first by adding one level tablespoonful of inoculant (about 6 g) to 30 millilitres (1 fluid oz) of water and stirring thoroughly. The inoculant slurry can then be added to the seed in a small glass jar or a can with a tightly fitting lid. By shaking the jar, the seed will be thoroughly coated with the inoculant particles.

Legume seeds should be inoculated just before planting. If any drying is necessary, it should always be done in the shade.

Other useful information on planning and putting on field demonstrations on inoculation is given in Tables 9 to 12. The amounts of inoculant and water needed for one cupful and for 30 ml of various seeds are given in Table 11.

Table 9. ROW SPACINGS, SEEDLING RATES AND PLANT POPULATION FOR VARIOUS LEGUME CROPS

	Between-row spacing (cm)	Within-row spacing (cm)	Seeds/m of row (approx.)	Plants/m of row after thinning	Plants/ hectare
Arachis hypogaea	60	5	25	20	333 333
Centrosema pubescens	20	20	[1] 10	5	250 000
Cicer arietinum	[2] 60 (30)	5	25	20	333 333
Desmodium uncinatum	20	20	[1] 10	5	250 000
Glycine max	60	3.7	35	27	450 000
Lens culinaris	30	5	25	20	666 666
Leucaena leucocephala	40	20	[1] 10	5	125 000
Medicago sativa	20	20	[1] 10	5	250 000
Phaseolus vulgaris	60	5	25	20	333 333
Stylosanthes guianensis	20	20	[1] 10	5	250 000
Vigna radiata	60	5	25	20	333 333
Vigna unguiculata	60	5	25	20	333 333

Note: The seed size of some legume species may vary by a factor of 2 or 3 between varieties. The values given here are for the larger-seeded varieties. If seed supply is a critical factor, then the weight of seed needed should be calculated for the specific batch of seed to be used.

[1] With a 20-cm row spacing, plant two seeds per hole and thin to one seed per hole (leucaena will later be thinned further to 40×80 cm). — [2] When grown as a dry season crop, the higher plant population should be used.

Table 10. INOCULATION METHODS AND RECOMMENDATIONS

Soil or planting conditions	Recommended method of inoculation
Small seeds: *Medicago* spp., *Trifolium* spp., *Desmodium* spp.	
Normal — good temperature, moisture, pH and seed-bed	Slurry — peat-base inoculant. 30 ml slurry/kg of seed
Acid soil, rough, poor seed-bed	Lime pellet. Inoculant slurry prepared with gum solution (gum arabic, or synthetic cellulose) 22 ml/kg seed. Apply double dose of inoculant
Normal — temperature, moisture, pH, seed-bed preparation good, heavy infestation of native ineffective rhizobia	Inoculant pellet. Apply 45 g inoculant/kg seed. Mix 15 g inoculant with 22 ml of gum arabic solution (45%) and apply to seed. Use the remaining 30 g inoculant to dry the seed immediately after applying the slurry
Hot, dry soil — seed-bed preparation good	See above
Seeds treated with toxic chemicals, other soil and planting conditions good	Plant without inoculation. Use soil inoculant after seedlings emerge. Use sod seeder and drill inoculant in rows 25 cm apart
Large seeds: *Pisum* spp., *Vicia* spp., *Glycine* spp., *Vigna* spp.	
Normal — good temperature, moisture, pH and seed-bed	Slurry. Apply peat-base inoculant using 4.4 g inoculant in 7 ml of water/kg of seed
Seed treated with toxic chemical dressings, other soil and planting conditions good	Use soil inoculant. Apply in seed furrow with seed at planting time
Hot, dry soil	Use soil or granular inoculant. Apply inoculant 2.5 to 5 cm below seed (or deeper if moist soil is deeper). Drill must be equipped with special feet for drilling deep
Fragile — seeds such as groundnut, *Arachis* sp.	Use soil or granular inoculant. Apply in furrow with seed

Table 10. INOCULATION METHODS AND RECOMMENDATIONS *(concluded)*

Soil or planting conditions	Recommended method of inoculation
Fields of non-nodulated legumes	
Small seeds sown broadcast	Sow granular inoculant with sod seeder, placing knives 25 cm apart and at a depth of 5 cm
Large or small seeds in rows	Sow granular or liquid inoculant in rows beside plants at a depth of 5-7 cm
Seedlings or cuttings	
Prevailing conditions	Prepare a slurry by mixing the inoculant with water: one volume inoculant to two volumes of water. Dip seedlings or cuttings in slurry and allow to soak for 1 minute or longer. Plant as soon as possible.

Table 11. AMOUNTS OF INOCULANT AND WATER NEEDED FOR VARIOUS LEGUME SEEDS FOR INOCULATION TRIALS

Kind of seed	Inoculant needed for: 1 cupful seed	Inoculant needed for: 30 ml seed	Water needed for: 1 cupful seed	Water needed for: 30 ml seed
	.... *grams*		.... *millilitres*	
Lucerne (alfalfa)	1.0	0.12	5.0	0.64
White clover	1.0	0.12	5.0	0.64
Vetch	1.0	0.12	1.6	0.20
Soybean	0.8	0.10	1.5	0.20
Cowpea	0.9	0.11	1.6	0.20
Horse bean	1.0	0.12	1.8	0.23

Table 12. RELATIONSHIP OF NUMBER OF SEEDS PER KILOGRAM TO THE SIZE AND APPROXIMATE NUMBER OF RHIZOBIA PER SEED FOR VARIOUS LEGUME SPECIES

Plant [1]	No. seeds 1 kg	Rhizobia No./seed	Kg seed 1 ha	inoculum g/ha
Vicia faba	1 000	22×10^5	87	384
Arachis hypogaea	2 000	11×10^5	44	194
Cicer arietinum	2 000	11×10^5	33	145
Dolichos lablab	3 000	7.3×10^5	27	119
Lupinus albus	3 000	7.3×10^5	110	484
Pisum arvense	6 000	3.7×10^5	98	431
Glycine max	9 000	2.4×10^5	65	286
Cajanus cajan	17 000	1.3×10^5	11	48
Lens esculenta	19 000	1.2×10^5	15	66
Vigna radiata	22 000	1×10^5	76	334
Cyanopsis tetragonoloba	44 000	5×10^4	38	167
Vigna aconitifolius	44 000	5×10^4	40	176
Vicia hirsuta	44 000	5×10^4	49	215
Trigonella foenum-graecum	50 000	4×10^4	33	145
Onobrychis viciifolia	66 000	3×10^4	38	167
Coronilla cretica	242 000	9×10^3	20	48
Ornithopus sativus	352 000	6×10^3	22	97
Desmodium intortum	440 000	5×10^3	11	48
Indigofera hirsuta	440 000	5×10^3	10	44
Medicago sativa	440 000	5×10^3	16	70
Trifolium incarnatum	605 000	3.6×10^3	18	79
Lotus pedunculatus	1 940 000	1×10^3	4	18
Trifolium agrarium	2 200 000	1×10^3	4	18
Lotononis bainesii	3 840 000	6×10^2	2	9

[1] From largest to smallest seeds in descending order.

Appendix. CONVERSION FACTORS

To convert column 1 into column 2, multiply by	Column 1	Column 2	To convert column 2 into column 1, multiply by
Length			
0.621	kilometre, km	mile, mi	1.609
1.094	metre, m	yard, yd	0.914
0.394	centimetre, cm	inch, in	2.54
Area			
0.386	kilometre^2, km^2	mile^2, mi^2	2.590
247.1	kilometre^2, km^2	acre, acre	0.00405
2.471	hectare, ha	acre, acre	0.405
Volume			
0.00973	metre^3, m^3	acre-inch	102.8
3.532	hectolitre, hl	cubic foot, ft^3	0.2832
2.838	hectolitre, hl	US bushel, bu	0.352
0.0284	litre	US bushel, bu	35.24
1.057	litre	US quart (liquid), qt	0.946
Mass			
1.102	tonne	ton (English)	0.9072
2.205	quintal, q	hundredweight, cwt (short)	0.454
2.205	kilogram, kg	pound, lb	0.454
0.035	gram, g	ounce (avdp), oz	28.35
Pressure			
14.22	kg/cm^2	lb/inch^2, lbf/psi ^{2}	0.0703
0.9678	kg/cm^2	atmospheres, atm	1.033
0.9869	bar	atmospheres, atm	1.013
Yield or rate			
0.446	tonne/hectare	ton (English)/acre	2.240
0.892	kg/ha	lb/acre	1.12
0.892	quintal/hectare	hundredweight/acre	1.12
0.87	hectolitre/ha, hl/ha	bu/acre	1.15

TIPO-LITO - SAGRAF - NAPOLI